Victor Pereira
Diego Fonseca Frade
Wemerton Luis Evangelista

Development of Training Methodology - Occupational Safety

Victor Pereira
Diego Fonseca Frade
Wemerton Luis Evangelista

Development of Training Methodology - Occupational Safety

Civil Engineering Works in the Municipality of Santa Luzia - MG

ScienciaScripts

Imprint

Any brand names and product names mentioned in this book are subject to trademark, brand or patent protection and are trademarks or registered trademarks of their respective holders. The use of brand names, product names, common names, trade names, product descriptions etc. even without a particular marking in this work is in no way to be construed to mean that such names may be regarded as unrestricted in respect of trademark and brand protection legislation and could thus be used by anyone.

Cover image: www.ingimage.com

This book is a translation from the original published under ISBN 978-613-9-72831-2.

Publisher:
Sciencia Scripts
is a trademark of
Dodo Books Indian Ocean Ltd. and OmniScriptum S.R.L publishing group

120 High Road, East Finchley, London, N2 9ED, United Kingdom
Str. Armeneasca 28/1, office 1, Chisinau MD-2012, Republic of Moldova, Europe
Printed at: see last page
ISBN: 978-620-7-89999-9

SUMMARY

The engineer in charge of a building site is not always aware of the occupational risks experienced by workers during their work. Occupational safety is the application of technical, educational and medical measures aimed at preventing and minimising accidents, making workers feel safe even in adverse conditions, acquiring prevention practices governed by the Regulatory Norms (NRs) established by the Ministry of Labour and Employment. The construction industry is characterised by its high turnover rate. This factor may be related to the adverse conditions encountered at work and the lack of professional qualifications. In addition, employees have no stability in the company, where they are employed during the construction period and then dismissed. The sector has poorly qualified labourers compared to other sectors, which means that low productivity results in products of lower quality than desired.

The aim of this study is to develop an effective training methodology to prevent/decrease accidents at work in the construction industry on construction sites in the greater São Benedito area, in the municipality of Santa Luzia/MG. In order to obtain the data, questionnaires were drawn up and submitted to site managers and workers, in which they answered questions relating to their social profiles, the organisation of the site, training and the availability of individual and collective protection equipment.

This research has shown that the occupational safety scenario in the construction industry has changed over the years. While in the early 2000s the concept of training and safety began to be demanded by the government, today it is imposed by the company. All those interviewed received basic safety training, first aid, handling and construction techniques during the working day, as well as receiving all the necessary PPE. This new measure adopted is fundamental to improving the safety of the worker, who often didn't have extensive experience in the area, carrying out their work in the way they liked best. But this is still not enough. The training offered is still not entirely effective.

Of those responsible for the work, 43 per cent said they had already witnessed an accident at work. With this in mind, a more effective training methodology was created to minimise accidents at work.

INDICE

CHAPTER 2

PROFILE OF CONSTRUCTION WORKERS IN SANTA LUZIA - MG

2.1. INTRODUCTION

Construction is one of the fastest-growing activities, both technologically and in terms of labour infrastructure for its workers. The development of cities and population growth have led to a greater need to develop this activity. The growing number of construction sites means that unskilled labour has to be hired, thus increasing the number of accidents at work.

The president of the Brazilian Association of Real Estate Developers (Abrainc), Rubens Menin, in his 2017 interview says: "There is no economic growth in Brazil without the growth of the construction industry. The construction sector accounts for 8 per cent of GDP. It's going to have a huge impact on the economy, on social development and, of course, on employment."

The profile of workers in the construction industry in the region of São Benedito, Santa Luzia, MG, is based on a comparison with the study of the profile of construction workers in Greater Florianópolis, Santa Catarina, whose main objective was to identify the professional, social and health situation, among others, with the need to use the data to show the reality experienced by workers in the formal sector in the region.

CHAPTER 3

WORKER PROFILE
3.1. NOTE

On the subject of "not informed", the workers did not answer the question on their questionnaires.

3.2. SEX

According to the field survey, there are a large number of male employees, 85% (graph 1), and a small number of females, 4%. Despite taking into account the high percentage of workers who did not state their gender, perhaps because they were afraid to make any kind of comment about the company, even though they had been informed that their identity would not be made available, the figure of 11% does not affect the research too much when the aim is to find out the predominant gender on the construction site.

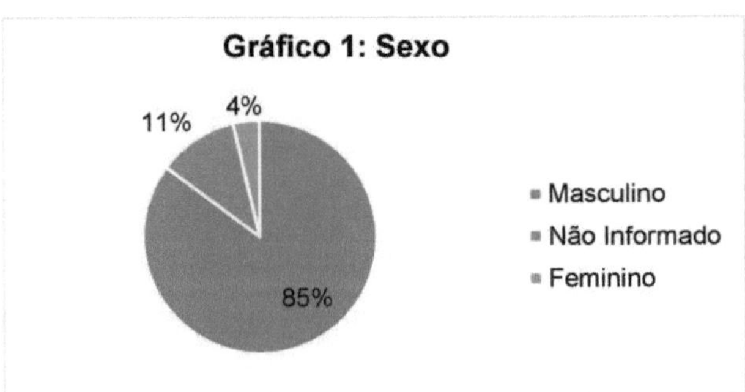

Graph 1: Sex
Source: author

The data obtained is relatively similar when compared to other research in the sector. According to data from the RAIS (Annual Social Information Report) in 2014, the number of male workers with a formal contract in greater Florianópolis represented 88% of the total contingent. While the number of female workers represented 12% of the total, a percentage not similar to that of this study, but relevant given the percentage of workers who did not inform their gender.

3.3. AGE

Among those interviewed, construction workers account for 17 per cent of young people (Graph 2). The highest percentage is in the 25 to 36 age group, which accounts for 44% of the total. Above the age of 36, there **is** a significant reduction in the number of workers, with a percentage of 16% aged between 37 and 48.

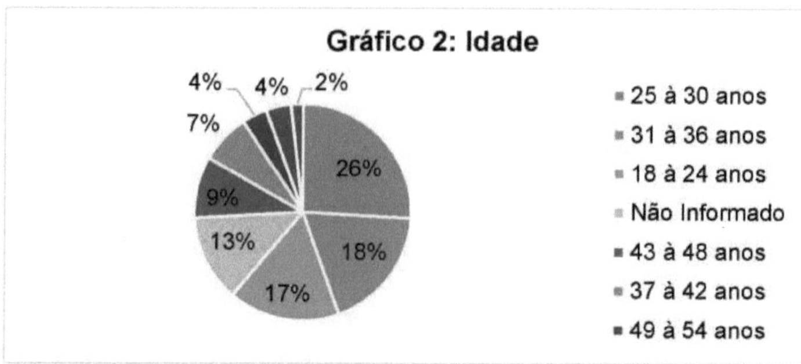

Graph 2: Age
Source: author

Expanding, comparing the age-related data from this work with that from the survey: "Sinduscon Seconci - Grande Florianópolis, Profile of the Construction Worker"; in 2013, the youngest workers (18 to 30 years old) were 55.60 per cent. In 2016, the same survey showed a drop to 37.90 per cent. As for this survey, in 2017, there is an approximate percentage of 43%; based on this, we can see changes that could show possible growth in the sector, especially for the younger generation, who can enter the labour market as opportunities grow.

3.4. RESIDENT CITY

According to the study, the graph shows that the workers interviewed are residents of Santa Luzia and Ribeirão das Neves, 30% and 31% respectively (graph 3), followed by Belo Horizonte, 17%. The number of interviewees who didn't inform us of their situation (13%) means that this percentage could influence the determination of which city would be the biggest employer of these resident workers. In relation to the other cities, we have Vespasiano (7%), followed by Betim (2% of the total number of workers).

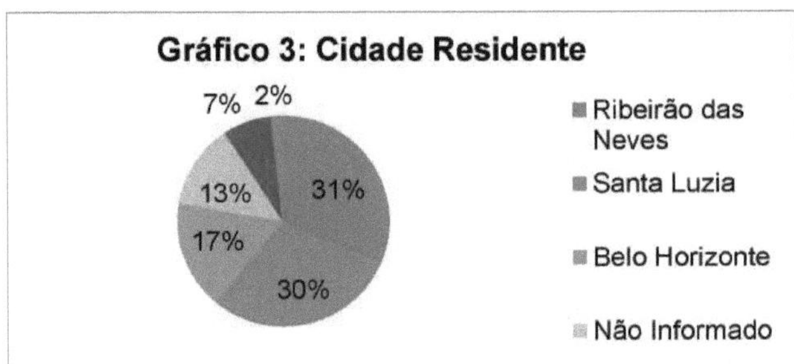

Graph 3: City of residence
Source: author

3.5. CIVIL STATUS

The information on marital status shows a higher concentration of married workers (48 per cent) and single workers (39 per cent). Few call themselves in a stable union, 11 per cent, and only 2 per cent are separated/widowed.

Compared to the 2016 Sinduscon Seconci Survey, 39.9% of respondents described themselves as married and 46.6% as single; there was a decrease in the number of married people and an increase in the number of single people, perhaps due to the growing number of young people entering the industry. With regard to stable unions, we have 9.3 per cent and separated/widowed approximately 4.3 per cent of workers.

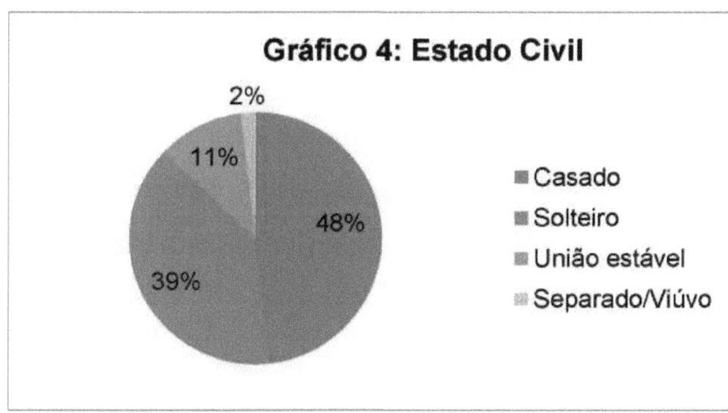

Graph 4: Marital status
Source: author

3.6. FAMILY INCOME

Workers' wage income is concentrated between one and three minimum wages (Graph 5), totalling 89%. One salary accounts for 33% of workers and two to three salaries for 56%. Above three salaries, a small number of workers can be seen, these being people with possible more valued positions in the construction industry, such as engineers, architects, master builders, among others. The percentage of workers earning between three and five salaries is 5 per cent, followed by 2 per cent of those earning more than five salaries and 4 per cent of those interviewed who did not state their situation.

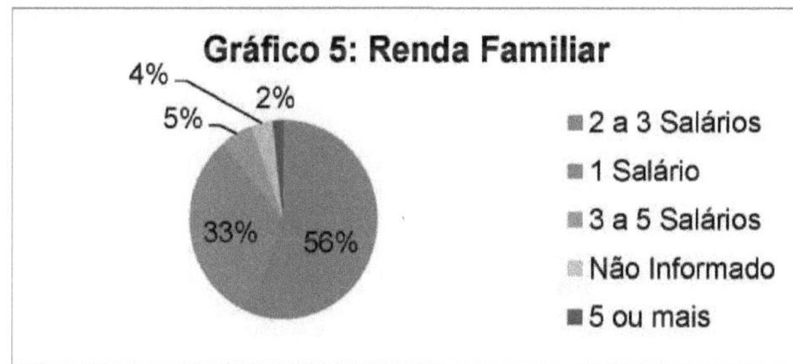

Graph 5: Family income
Source: author

3.7. SCHOOL EDUCATION

If you look at the graph with the survey data, Graph 6, you can see that the workers have a low level of education, with the majority not having completed primary school, 42% of the workers. Based on the completion of secondary school, 26% have completed it and only 2% of those interviewed have gone on to higher education, but none of them have completed it. Regarding the completion of a technical course, 2% of the workers entered and completed the course.

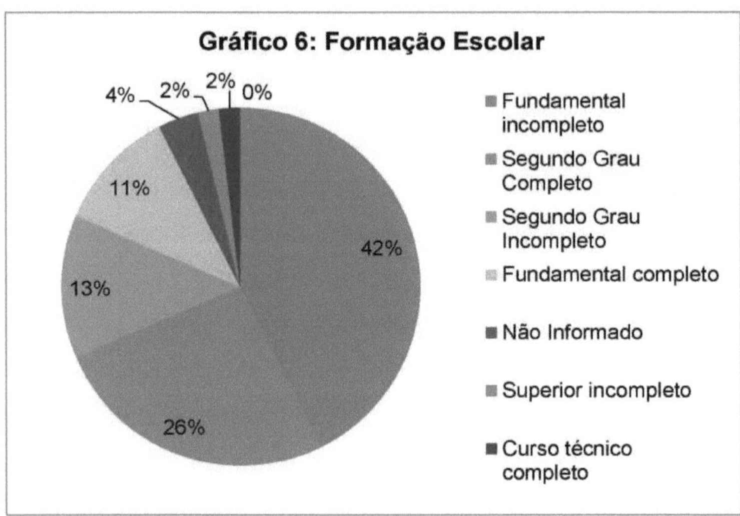

Graph 6: Schooling

Source: author

Comparing the data with the 2015 Sinduscon Greater Florianópolis survey, 33.8% of workers have not yet completed primary school and only 13.2% have completed secondary school, an increase of over 10% in two years. With regard to higher and technical education, 5.1 per cent have started higher education and 2.4 per cent have completed it, unlike what is shown in this study, which in this case, none of the workers have completed it yet.

3.8. FUNCTION PERFORMED

Looking at the jobs performed by workers (Graph 7), it can be seen that the jobs with the highest density of professionals are those performed by servants (30%), followed by bricklayers (28%) and carpenters (7%). Compared to the

2016 Sinduscon Seconci survey, there has been a considerable increase in the number of servants, with 18.5 per cent in this survey, while bricklayers accounted for 20.6 per cent, followed by carpenters with 14.2 per cent, confirming the idea that opportunities are growing in the construction industry due to the greater number of these professionals in the field.

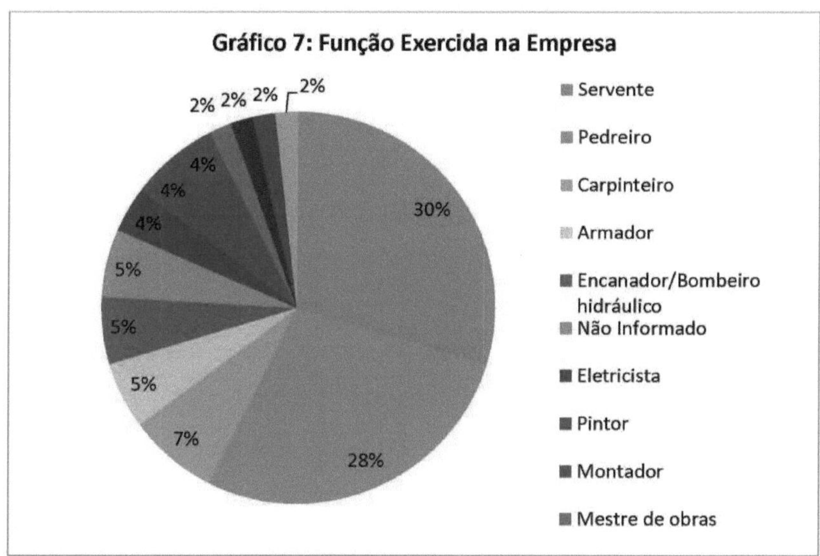

Graph 7: Position held in the company
Source: author

Looking at the jobs with the lowest density of employees, we have watchmen, master builders, fitters, painters and foremen,

Auxiliary, Electrician, Plumber totalling 30%, based on a comparison with the professional data from Sinduscon 2016, in which these functions are classified as "Other", the percentage totals 39.1% of the data.

In relation to the interviewees who did not inform us of their situation, we have 5% of the percentage of workers, a figure which could influence the function of greater professional density.

3.9. LENGTH OF SERVICE

Most of the interviewees have been working in the industry for less than 10 years (Graph 8), with 35% of the workers having been working for less than five years. This may demonstrate the tendency towards a high turnover of workers, due to the low level of specialisation required for most positions. However, it is still possible to highlight a considerable number of workers who have been with the company for more than 10 years, 20 per cent. Compared to the 2016 Sinduscon Seconci survey, workers with less than 10 years' experience account for 47.8% of employees, compared to 67% in this study, confirming the high turnover in the construction industry, which is often due to business owners' fear of investing in qualification courses, due to the various specialities in the construction industry, which require a lot of specialisation and investments more focused on particular qualifications rather than a complete course in which the worker can learn the basics of each function.

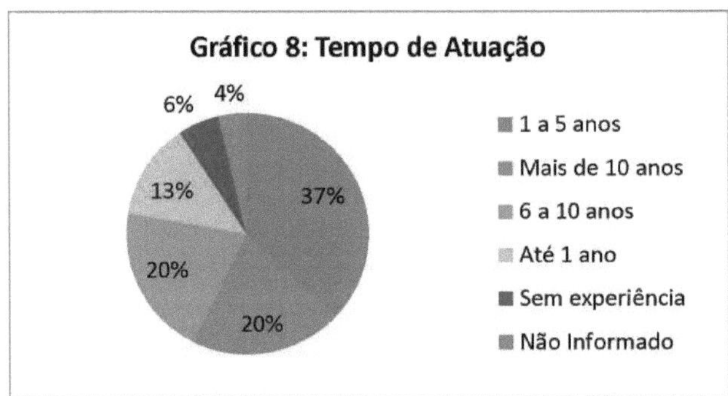

Graph 8: Length of service
Source: author

According to a 2016 Sinduscon Seconci survey, construction companies have difficulty finding qualified workers for all job roles. Faced with the problem of labour shortages, the mechanism used by companies is the qualification of professionals on site.

3.10. HOW YOU LEARNT THE FUNCTION

There is a predominance of people who learnt the activities of their profession in their own work environment (graph 9), with 83% of the workers interviewed. In addition, there is a low percentage of employees who have taken vocational courses (only 5%), a percentage very similar to the Sinduscon survey, which shows 7.6%. With regard to other types of learning, we have those who learnt in the family with 4%, a different percentage to Sinduscon, which shows 12.3%. With regard to higher education, in both surveys the data was no higher than 5% of the total percentage of employees.

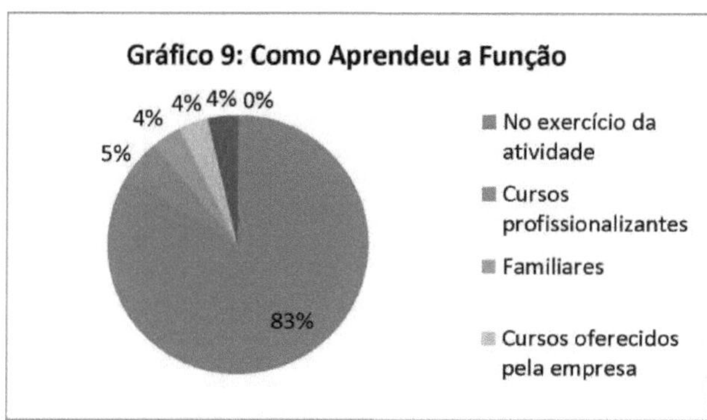

Graph 9: How you learnt the function
Source: author

3.11. TRAINING

Turning to safety issues, during our visits we noticed that the construction sites were kept clean and organised. They all had a storeroom, a rest area, toilets, an office and a storage room. The materials used in construction were stored in places that did not impede the passage of pedestrians and vehicles. At all the sites visited, there was a control system for people entering the site. Everyone was properly equipped with the necessary PPE, such as helmets, boots and gloves.

Training is fundamental for carrying out tasks on a building site. With the ever-increasing demands for better working conditions, companies organise training courses (chart 15) for all employees. They are taught by employees in the field. The workload varies, with some training sessions lasting less than 10 hours and others up to 20 hours, but they are always held during the working day, which means that employees have no excuse not to take part.

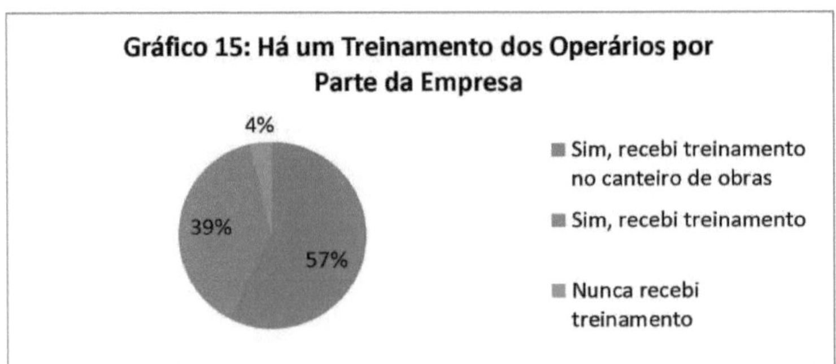

Graph 15: The Company Trains Workers
Source: author

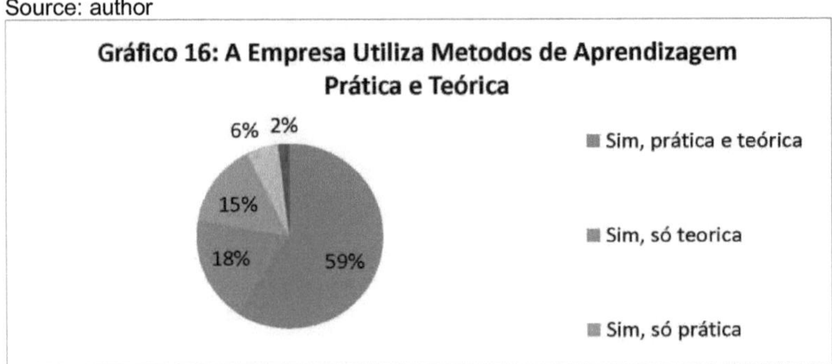

Graph 16: The Company Uses Practical and Theoretical Learning Methods

Source: author

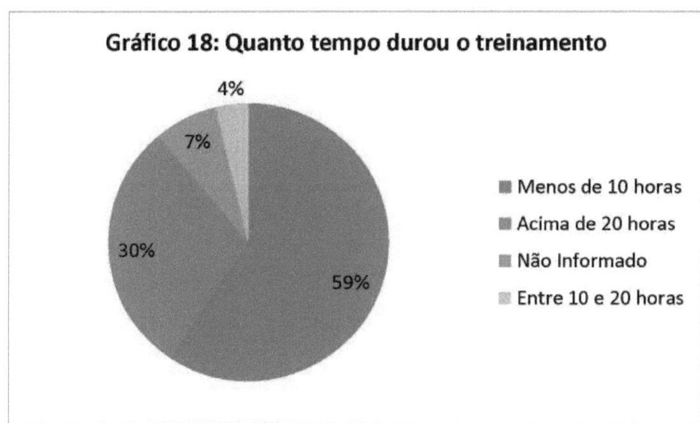

Graph 18: How long the training lasted
Source: author

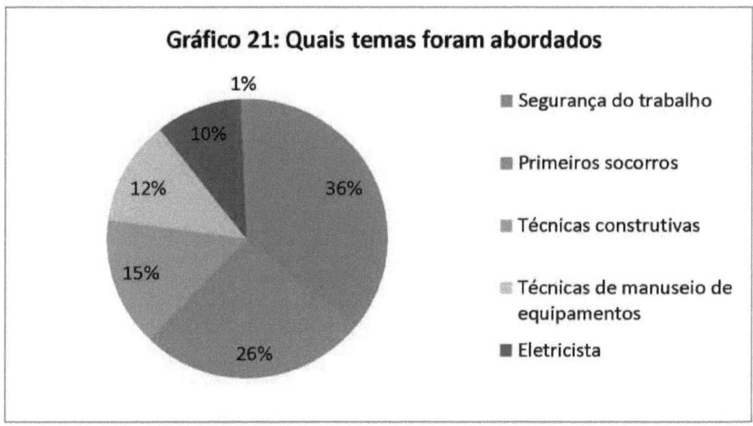

Graph 21: What topics were covered
Source: author

CHAPTER 4

TRAINING FOR CONSTRUCTION WORKERS: BASIC NOTIONS OF EQUIPMENT AND FIRST AID

4.1. HOW CAN TRAINING HELP MY COMPANY?

The answer to this question is simple: with motivated employees who are aware of the activities they need to carry out, they will work better and the quality of service will increase.

With training, it's also possible to discover new talents, without having to invest again in hiring other professionals, so that the company and employees benefit from these newly discovered skills.

4.2. MOTIVATING THE TEAM

Keeping employees motivated is essential to maintaining the health of the company. Creating a practice to introduce the company to employees, showing the principles, positive points and what needs to be improved is essential. The manager should also take the time to be closer to the team, showing that the leader is with them.

Using feedback as a useful tool is an individual way of making it clear to the employee which points need to be improved, without them suffering any consequences. This tool should be applied on an individual basis, so that the employee feels comfortable raising the issue. There's nothing better than the team involved in the problems to make suggestions on how to solve them.

This feedback should go both ways; the company shouldn't be afraid to praise and give credit, after all everyone appreciates their efforts being recognised. You just have to be careful not to treat one member differently, as this could be interpreted differently by the others, making them feel undervalued and demotivated.

Celebrate achievements. When the team achieves goals or solves problems, it is necessary to celebrate, something that is feasible, such as a snack. In the long term, these attitudes can make a difference, as they increase productivity and competitiveness.

CHAPTER 5

TRAINING

- Behaviour in the company
- Handling basic tools
- How to act in the event of an accident

5.1. GENERAL OBJECTIVE

Train workers how to behave, handle tools and carry out their activities on the construction site.

5.2. HOURS

16 Hours

CHAPTER 6

METHODOLOGY

For the training, a space will be needed to accommodate the workers within the construction site itself, with the possibility of projecting images. A website is also available with the content presented during the training to answer questions.

The presentation of the materials and the construction site will take place in two stages. In the first stage, employees will be shown the written part with quick phrases of what it is and how to use it, along with images. In the second stage, the equipment will be presented visually, leaving physical and visual contact, showing its use, care and examples of misuse.

The site will first be presented indoors, showing the main points, such as the offices and warehouse, signposts, organisation and control of entry and exit. After the presentation, the employees will be taken to the points where they will be visually shown the signs and where the materials will be stored. This will be a time for relaxation, where in addition to getting to know the workplace, they will also be able to get to know their fellow workers better, including the engineer(s) in charge.

Figure 1: Training steps.

The training follows these steps:

- A date and time are set for the training session and everyone is notified, so it's important that everyone attends.

- The training provider is defined as the site safety technician, together with the engineer in charge whenever possible.

- The location of the training is defined, in a covered place within the construction site, usually in the canteen.

- The participants in the training are gathered together, the trainer and the employees to be trained, at the location where the training will take place. The training items to be presented are presented visually, using images, videos and the equipment they will be using, and commented on one by one.

  - The construction site is reconnoitred, showing the main spaces: offices, canteen, storeroom, toilets, rest area, free access areas and material storage areas.

  - The main signs on the construction site are shown, with each one commented on.

  - It's extremely important for employees to have a voice and to be able to share their experiences, propose the best way to carry out a task and ask questions when they arise.

CHAPTER 7

PROCESSES AND TECHNIQUES
7.1. ORGANISING THE CONSTRUCTION SITE

Keeping materials in the right places at all times, leaving circulation spaces free, reduces the risk of risk situations. The construction site must be kept clean in order to avoid trips, slips and reduced travelling speed. Debris and leftover materials should be continually removed from where they are piling up, and properly placed in a more appropriate place for disposal or later reuse. Eliminating dust also helps prevent accidents.

7.2. HAZARDOUS MATERIALS

Toxic, corrosive and flammable materials should be stored in an appropriate and well-protected place. Their packaging must be labelled to indicate the quality of these substances.

7.3. SIGNALLING

With the intense movement during the working day within the site, signalling must be respected.

According to Regulatory Standard 26 (NR 26), safety signalling must identify safety equipment, delimit areas and comply with local technical standards. Each colour is intended for a different purpose, as shown below.

7.3.1. RED

They are used to indicate fire protection and fire-fighting equipment and apparatus, such as fire alarm boxes, hydrants, fire extinguishers and their location, and fire hoses.

7.3.2. YELLOW

This colour is used to indicate "Caution", showing places with edges of floor openings, lift doors that close vertically, strips on the entrance floor of lifts and loading platforms, cabs, buckets, cranes, excavators and suspended equipment that could pose a risk.

7.3.3 WORDS OF WARNING

DANGER: indicates high risk

CAUTION: indicates medium risk

WARNING: indicates slight risk

CHAPTER 8

PERSONAL PROTECTIVE EQUIPMENT

Safety equipment is indispensable during the working day and is the responsibility of the worker,

keep it preserved and inform the employer of any alteration that may make it impossible to use.

The company is obliged to provide its employees, free of charge, with the necessary PPE in good condition and working order.

8.1. THE EMPLOYER IS RESPONSIBLE FOR THE PPE

- Demand its use;
- Use it appropriately according to the risk of each activity;
- Guiding workers on its use and conservation;
- Replace immediately if damaged;
- Take responsibility for regular maintenance.

8.2. IT'S UP TO THE EMPLOYEE TO WEAR THE PPE

- Require the employer to provide it;
- Use it at all times while carrying out your work activities;
- Take responsibility for conservation and safekeeping;
- Inform your employer of any changes that make it unsuitable for use;
- Comply with the employer's instructions on proper use.

8.3. HELMET (skull protection)

The main piece of protective equipment, it has a wide application, protecting the individual from objects that could fall from different levels. This object must remain unchanged, without any modification by the worker.

The main function of this equipment is to protect the worker's skull, since in the construction industry there is work at different levels and objects can be accidentally thrown at an employee. It also provides some kind of protection against electric shocks, burns and liquid splashes.

Figure 2: skull protection

8.3.1. USE:

For proper use, objects should never be placed between the suspension and the hull, as this space is necessary to absorb impacts;

-Always check   before use that there are no cracks, holes or changes to its shape and colour;

-The   flap must remain facing forwards and must not be tilted;

-When you   need to attach an accessory, follow the instructions on the accessory correctly, and make sure it is from the same manufacturer, so that it fits perfectly;

-Do not affix adhesives or apply chemical substances, as this may reduce the equipment's resistance;

-Wash it   periodically with warm water and neutral soap using a sponge.

8.4. FACE AND EYE PROTECTION

Face shields provide protection against objects that can be thrown, burns and the action of heat and radiation.

8.4.1. WELDER'S MASK

This equipment is specifically for welders. Its function is to protect the face against radiation from the welding operation, splashes of molten metal and welding sparks.

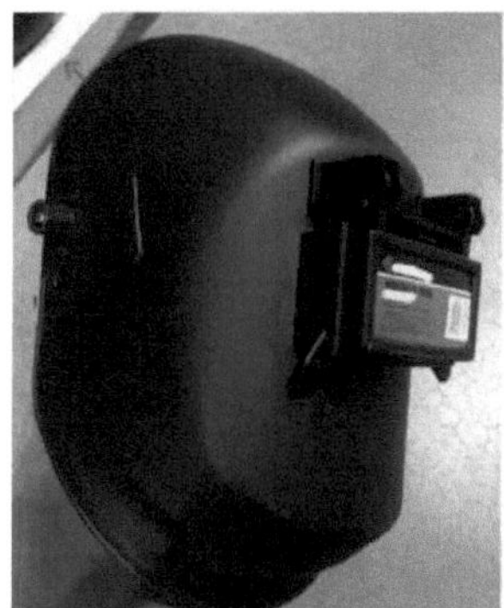
Figure 3: welder's mask

8.4.2 EYE PROTECTION

Figure 4: safety goggles

Eye protection is very important for accident prevention. These organs must be protected from sparks, dust, splashes of mortar, cement, splinters and all objects that can damage vision.

8.5. RESPIRATORY PROTECTION

Within the construction site, there is always the inhalation of dust caused by various products. Therefore, respiratory protection equipment (RPE) is one of the most important PPEs for protecting workers' airways.

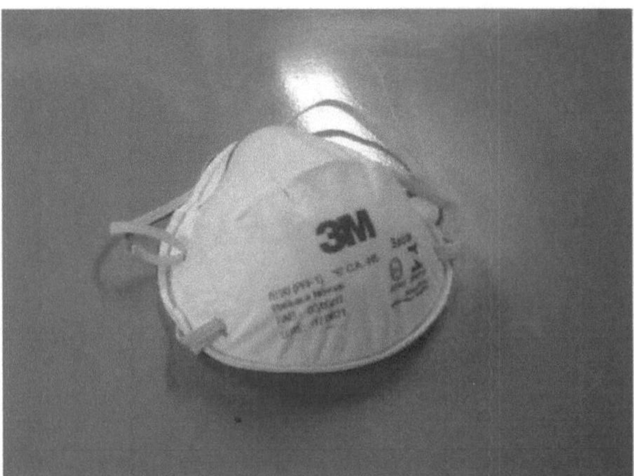
Figure 5: Respiratory protection mask

The mask must provide a perfect seal in the areas where it comes into contact with the face. It must be easy to use, comfortable and not compromise communication.

For good conservation, it should be cleaned after use with warm water and neutral soap.

8.6. HEARING PROTECTION

Work in places where exposure to levels above the tolerance limits established in NR 15, there must be control of the source or the environment, to reduce noise from the sources. If this is not possible, individual hearing protection methods should be adopted.

Among the protectors, two can be adopted:

8.6.1. INSERT PROTECTORS

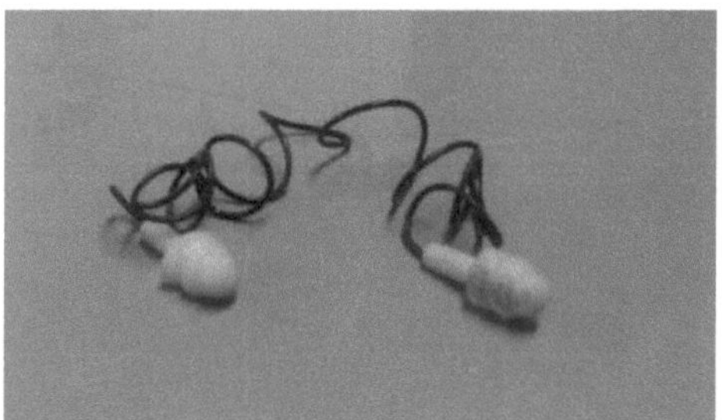

Figure 6: Insert earplugs

They are mouldable, usually made from polymers or silicone. They should be sanitised periodically to avoid infection.

8.6.2. SHELL PROTECTORS

The shell-type protector is made up of two elliptical cavities held together by a rod. This equipment is not recommended for people who wear glasses or have long hair, as it is suitable for smooth, soft surfaces.

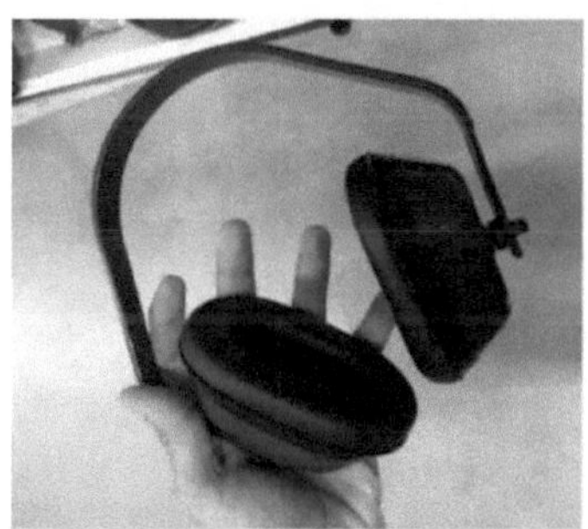

Figure 7: Shell-type ear protection

8.6.3 SOME CONSIDERATIONS

-The   choice of protectors must be based on the user's comfort;

-Hygienisation must be carried out in accordance with the manufacturer's instructions;

-When choosing, consider: sealing, efficiency, comfort, ease of use and compatibility with other PPE.

-When damaged, they must be replaced.

8.7   PROTECTION FOR UPPER LIMBS

The upper limbs are more prone to injury during work activities. Without a doubt, the hands are the most likely to suffer damage, and safety gloves are the best protection.

Figure 8: Protective gloves

To choose the most suitable material, you need to visualise what protection the employee needs (hands, forearm, arm); the chemical composition of the substances handled; humidity and temperature conditions.

You should be aware of which jobs require gloves and which are contraindicated, such as on certain machines, when they can create unsafe situations and cause harm to the worker.

8.8. LOWER LIMB PROTECTION

The protection of these limbs is necessary due to the conditions or aggressive agents at work. They protect the lower limbs against falling materials, electrical discharges, burns caused by chemicals and sharp objects (Neto, 2014).

There are two main types of safety footwear. Boots with toecaps are safety shoes that have a steel toe cap on the front to better protect the feet. There are also boots without toecaps, recommended for electricians.

8.9. FALL PROTECTION

Collective protection measures against falls from a height are mandatory and a priority. Where these measures are not possible, a parachute-type safety harness should be used.

The safety harness must be used when working at a height of more than 2 metres, when there is a risk of the worker falling. This belt must be fitted with a fall arrest device, connected to an independent safety cable and not attached to scaffolding, for example.

The right type of cable to use is steel or synthetic fibre, provided it is in perfect condition.

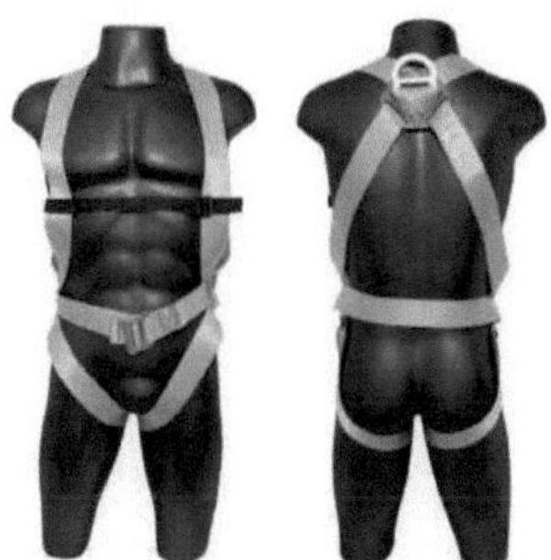

Figure 9: Fall arrest belt

8.9.1. SOME CONSIDERATIONS

- Before each job, check the condition of the equipment;
- Use one belt per user for the entire activity, so that adjustments don't have to be made frequently;
- If the belt falls, it must be replaced;
- Take care that the belt does not suffer cuts, burns or other injuries;
- Do not modify the belt, it must remain in its original form.

CHAPTER 9

USE OF BASIC TOOLS
9.1. WALL PLUMB BOB

The plumb bob is used to check the plumbness of walls in the masonry or plastering phase and also to check the plumbness of plastering masters. The wooden part should be supported on the wall so that the metal part is tangential to the wall. If it is leaning against or too far away, the plumb line is incorrect.

9.2. IMPACT DRILL

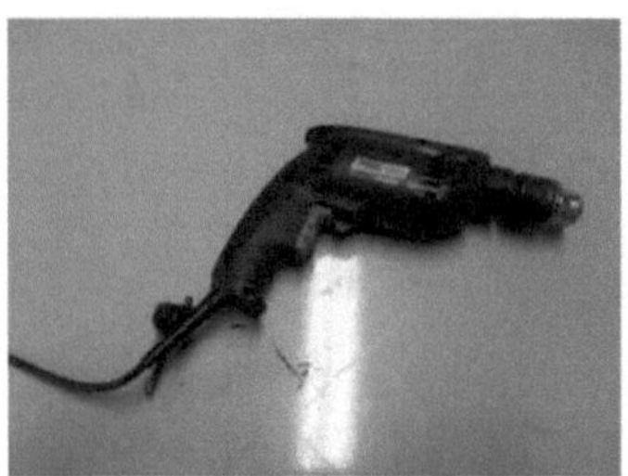

Figure 10: Impact drill

To use this equipment, first choose the drill bit you want. With the drill switched off, turn the chuck (base of the bit) so that the jaws of the bit are open and fit the bit all the way in. Turn the chuck again to secure.

The chuck has 3 holes around it, with its own spanner, which comes with the drill and is T-shaped, fit it into one hole at a time and tighten, allowing the drill bit to sit firmly.

Plug it in and test it. Check if the drill has any lateral vibration. If it does, replace the bit again.

Hold the handle where the trigger is with your hand, it will be responsible for activating the drill by squeezing the trigger. The other hand goes in a U-shape to support the front of the machine's base, taking care not to cover all the drill's air vents.

For your safety, wear a glove on the hand that activates the trigger, a mask to avoid inhaling dust, ear plugs to minimise the noise of the drill and glasses to avoid getting dust in your eyes.

9.3. MARBLE SAW (MAKITA)

Figure 11: Marble saw

The marble saw has two settings, cutting and angle. To adjust the angle, simply adjust it from the angle part of 90 degrees down by loosening the adjustment screw and tightening it again. This screw is on the front of the tool. To adjust the depth, simply loosen or tighten the screw next to the hand grip.

The saw comes with a water cooling system for more difficult cuts, preventing the disc from overheating and also avoiding the dust generated by the cut. You should take note of which disc you are using, as there is a specific disc for wet and dry cuts.

There is a locking button next to the trigger. This allows the saw to operate without having to keep pressing the trigger. To use it, simply press the trigger and then push the lock button. To switch it off, press the trigger.

9.4. POLICORTE

This tool is used for cutting steel, iron, aluminium, profiles and tubes. It must *only be operated by a fitter and the site must be identified with the photo of the professional authorised to carry out the operation.*

When handling it, you must observe the direction of rotation of the motor, which must be anti-clockwise, and check that the cutting disc is securely fastened. Secure the material, start the engine and gradually lower the disc until cutting begins, keeping the speed constant. For the safety of the operator and people nearby, always secure the material so that it is not thrown. Only people using the machine should be allowed on the site. The bolts should be checked and tightened daily and the machine lubricated weekly. Whenever the machine is operated, the operator should wear gloves, goggles, a mask and noise mufflers.

Figure 12: Poly-cut saw

9.5. CIRCULAR SAW BENCH

The circular saw must be located in a place where it is easy to remove the materials, signposted with yellow strips on the floor and the carpenter responsible must be indicated with a photo of the professional.

Fine debris such as sawdust and dust should be removed daily,

When operating the machine, the worker must wear gloves, a mask, ear protection and goggles.

Figure 13: Circular saw bench

Figure 14: Circular saw bench

9.6. CONCRETE SAW (Clipper)

Used for cutting concrete, cutting expansion joints in concrete floors and cutting Interlocking floors to finish the floor.

Figure 15: Clipper saw

Before starting work, you should check the direction of rotation of the motor and make sure that the reservoir contains enough water for the operation, as this is responsible for cooling the cutting disc.

For safety, the operator must wear boots, a mask, gloves and goggles. Check the screws and the saw daily and always transport it with the engine facing upwards. The cutting disc must be suitable for the floor and must always be cleaned after use.

9.7. SANDERS

Its function is to grind steel or concrete surfaces. They also do rough grinding.

9.7.1. USE:

- Depending on the type of material to be sanded, you will need a particular type of sanding paper. It is placed at the front.

  -Pass the sander in the direction of the wood grain. Hold it firmly with both hands.

-When finished, it should be cleaned with a brush.

-When handling it, the operator must wear gloves, goggles, a mask and ear protection.

Figure 16: Sander

9.8. COMPACTOR TOAD

The soil compactor is a versatile piece of equipment and its operation is reasonably simple, but it does require some experience, strength and firmness in the hands. Ideally, this type of machine should always be operated by someone who is qualified. Compactors are used to compact granular and mixed soils, and are often widely used to form bases and foundations for floors.

Figure 17: Toad compactor

Compactors are heavy machines for manual operation, usually weighing between 50kg and 60kg and with very powerful petrol engines.

9.8.1. USE:

-Do not use the equipment indoors;

-Do not smoke or generate sparks near the machine, as it uses flammable liquids and emits gases;

-Avoid compacting areas near ditches, ravines and slopes that pose a risk of collapse or landslides;

-Keep hands and feet away from moving parts of the equipment;

-Prepare the ground, removing anything that might hinder the operation;

-Position the compactor on level ground, preventing it from falling or slipping out of the operator's hands;

-Be careful not to compact sharp points, which could damage the accordion;

-Check the percussion oil, its level and top up with petrol (these procedures must be carried out with the compactor switched off);

-Avoid driving over tarmac, blocks or cobblestones.

9.8.2. TRANSPORT

-Empty the petrol tank;

-Secure	the equipment to prevent it from moving during transport;

-When hoisting or lifting the machine, attach lashings only in the places indicated;

-Stand the equipment upright in its working position;

- Check that your accessories are being taken with you and take special care with wires and cables.

9.9. CERAMIC CUTTER

Ceramic cutters generally cut at two angles, 45° and 90°. The scriber only marks the piece, do not force it to cut, this function is done by the separator in conjunction with the lever. The equipment must always be lubricated, especially the scriber which must slide easily (Dutra, 2013).

Use gloves and goggles to prevent small splinters from reaching your eyes.

9.9.1. USE

-Measure	the cut, at 45° or 90°;

-Fit	and secure the workpiece in the cutter so that it does not move;

-Push	the lever forwards to scratch the coating without applying force (twice is enough);

-Place	the workpiece on the separator, positioning the scratch in the centre;

-Carefully lower the lever.

Figure 18: Ceramic cutter

9.10.BOOTH

Figure 19: Concrete mixer

You should always check that the mixer is stable considering its weight and the continuous movement it makes when mixing the concrete, with no vertical movements. Its angle should be 45° during mixing.

Cement should never be poured before water so that it doesn't stick to the walls and damage the concrete's strength.

It is recommended to wear safety goggles, hearing protection, gloves and boots when using it.

9.7.2.   . USE

- It is recommended to add 1/3 of the water and 5 padioles of area to the mixer already in operation;

- Switch off the equipment and pour all the cement into the machine;
- Then, with the equipment switched back on, add the rest of the sand trays;
- Add as much water as you need to get the mixture just right;
- Finally, mix until completely homogeneous, and with the equipment still switched on, pour the mixture into the container to start working;
- Use the pedal and pinion of the steering wheel to pour the concrete.

CHAPTER 10

MENTORS FOR NEWCOMERS

A mentor must have experience linked to the demand. They don't have to have done very well in their role, but they can have "scars", i.e. they have learnt from mistakes what the situation they are going to teach is like (Gasparini, 2017). Working with someone who can help them avoid the mistakes they've experienced or seen someone else make gives them valuable information.

The construction industry is home to many workers who are just starting out in their careers. It can be seen that a large percentage of them have little experience, with 18 per cent having up to a year's experience or having just entered the field. This leads to unsafe situations arising during work activities.

Appointing mentors for these workers is a good way of reducing these situations. Thus, a worker who already has experience in the field will be responsible for monitoring and developing the newcomer.

As soon as the worker receives the name of his mentor, he must accompany him in his activities. After the training offered by the company, they should initially observe and help out as requested. The mentor must explain the activity they are carrying out, detailing step by step how it should be done. They will also be responsible for answering any questions that arise about the processes. Secondly, the worker must begin to implement what they have been taught, supervised by the mentor.

At this point, if they are involved in a risky activity, the newcomer should feel confident and be motivated by their mentor. If they don't feel confident about the activity, they should ask for help and be guided as far as they can. If they don't feel able to complete the activity, they should take a break and let their mentor take over.

CHAPTER 11

SPECIFIC TRAINING

11.1. FURNITURE

Masonry consists of the assembly of blocks interconnected by a mass of cement. It is done with two basic coordinates, with a vertical line levelling the slope and a horizontal line levelling the length. Well-executed masonry is perfectly flat vertically, requiring little coating thickness (KOPSCHITZ, 2011).

11.1.1. EXECUTION

1°). Mark out the walls based on the building's (architectural) floor plan, fix a tile with mortar and lay the first row using a line, square, plumb and level;

2°). Execute all the rows following a levelled line clamped between two guide plumb bobs, using the plumb bob and laying the bricks in a "mata-junta" system (mismatched vertical joint);

3°). To fill the gap at the top of the wall, before it meets the concrete beam, the masonry is tightened, which consists of laying solid bricks laid at an angle with weak mortar.

11.1.1.1. CHAPISCO (1ª layer)

- Thin layer of cement and washed coarse sand 1:4 to increase adhesion with the plaster (2ª layer).
- Apply the mixture using a trowel so that it is well spread out.

11.1.1.2. EMBOWMENT (2nd COAT)

- Apply only after the render has dried, with the electrical, water and sewage installations already in place;
- Insert master guides at the ends of the walls so that the trowel is in a horizontal plane;

- Apply a layer between 1.0 and 2.5 cm thick by spreading the mortar and smoothing it with a trowel, following the guidelines.

11.1.1.3.   REBOCO (3ª Layer)

- Apply after the plaster has dried;
- Apply a layer 0.5 cm thick, hitting the surface with an aluminium ruler.

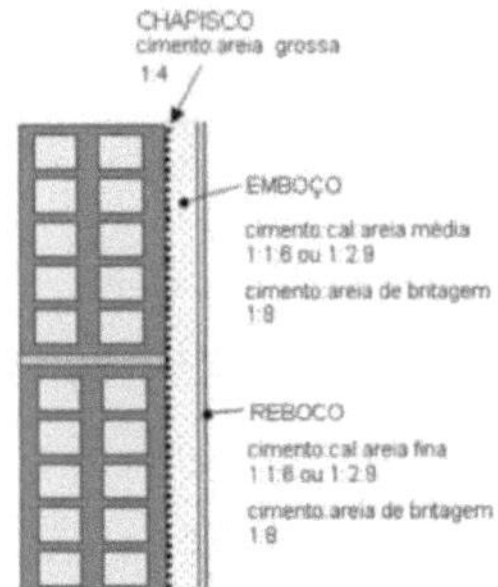

Figure 20: masonry

11.2. PRECAST

The prefabricated panels arrive completely ready to be installed. They are made by connecting metal *inserts that are* fixed to both the parts and the building structure. Assembly alignments, tolerances and transport must be carried out in such a way as not to compromise the structure.

The joints between the panels must be as watertight as possible and must have enough space to accommodate the movement of the panels, complying with the maximum and minimum widths determined by the panel movement calculations presented in the project. These gaps must be sealed with the appropriate material to prevent infiltration.

11.3. HYDRAULIC INSTALLATIONS

No cuts (either vertical or horizontal) are allowed for the passage of pipework. Some electrical pipes will pass through vertical holes in the blocks, while the water and sewage pipes will pass through a false wall (shaft), strategically positioned in the project, usually behind the place where the shower will be inserted (KOPSCHITZ, 2011).

Figure 21: shaft

Attention must be paid to the correct positioning of openings for the passage of pipework. Taps, valves and other fittings must be installed in accordance with the project, observing the heights and horizontal distances. A test should be carried out before coating the walls, plugging the outlets with fittings and filling the water tank so that the pipework is filled.

11.4. ELECTRICAL INSTALLATIONS

This is the stage of installing electrical conduits, switches, boxes and other installations required for electricity. It starts with the temporary power connection at the building site, goes through the installation of pipes and built-in boxes, and ends with the passage through the masonry to the final connections (sockets and switches).

The diameters to be installed must correspond to those specified in the project.

The work must be carried out by a trained professional, wearing all the PPE that guarantees their safety and with the power switched off (KOPSCHITZ, 2011).

11.4.1. IMPLEMENTATION
11.4.1.1.   ELECTRODUCTS

Install the pipes for the horizontal sections in the slabs before concreting and the vertical sections in the walls after the masonry has been completed;

Take care with flexible pipes so that they are not damaged during concreting.

11.4.1.2.   BOXES

- It must be installed taking into account the height of the finished floor;
- Switches and sockets should be 1.10 to 1.40 metres apart;
- Low sockets should be 0.30 metres away;
- Showers should be between 1.90 and 2.10 metres.

11.5. CARPENTRY

Woodwork must be fixed by a joiner. The trim should be fixed after the walls have been finished. Doors are inserted and held in place by metal hinges on the trim. They must open and close perfectly, without catching.

1.6. .FRAMES

The frames must be inserted in such a way as to prevent rainwater from seeping through to the outside of the building, and the gaps between the frame and the façade must be sealed with a flexible putty (silicone) (KOPSCHITZ, 2011).

1.7. .CONTRAPS

Layer of mortar applied approximately 2 to 3 cm thick on the slab.

1.7.1. . EXECUTION

- Apply copings to define the execution plan by removing the level of the environment;

- Spread the mortar following the slats;
- Finish with a trowel.

1.8. CERAMIC FLOOR
1.8.1. . EXECUTION

- Clean the subfloor thoroughly;
- Spread the mortar on the subfloor with a steel-toothed trowel in small sections;
- Apply to the floor with light strokes;
- Leave joints of approximately 2.0 mm;
- Apply skirting boards in a uniform size, following the laying of the floors;
- Remove the spacer when the mortar has set sufficiently;
- Apply the grout, filling the gaps between the tiles evenly.

11.8.2. REMARKS

Some ceramic tiles must be pre-wetted, pay attention to the manufacturer's instructions (KOPSCHITZ, 2011).

11.9. BLUE

- Applied after the floor has been finished;
- Make sure the floor is cut so that both parts can be used;
- They should be applied in a floor-to-ceiling direction, calculating the height of the rows so that the last piece on top is whole;
- Apply the industrialised adhesive mortar with a steel-toothed trowel over an area of approximately 1 m^2 , so that the entire surface of the tile is filled;
- Glue the tiles together leaving a joint about 2.0 mm wide, tapping lightly;
- Place a spacer between them;
- Remove the spacer when the mortar has set sufficiently;
- Apply the grout, filling the spaces between the tiles evenly (KOPSCHITZ, 2011).

11.9.1. OBSERVATIONS

- The wall must be free of moisture;
- The plaster must be semi-damp before gluing;
- Lightly moisten the tiles before laying;
- Avoid stressing the wall after application.

1.10. . PAINTING

One of the last stages of the work must be done after preparing the surface, which must have been sanded and cleaned, free from dust, oils or any kind of material.

It can be applied using brushes and rollers. Brushes are used to finish off corners and edges that the roller was unable to reach (KOPSCHITZ, 2011).

1.10.1. . EXECUTION

- Sand and clean the walls;
- Dissolve the paint in drinking water as specified by the manufacturer;
- Apply two or three coats to the smooth wall using a roller;
- Finish the corners and edges with brushes so that they match the colour of the wall already painted.

CHAPTER 12

HOW TO ACT IN THE EVENT OF AN ACCIDENT

In cases of fainting, convulsions, fractures, dislocations, cuts, punctures and poisoning, call the Mobile Emergency Assistance Service (SAMU) on 192, and in cases of cave-ins call the Fire Brigade on 193.

How to act in a given situation is described below:

12.1. FIRE BURNS

If the person is on fire, make them roll on the floor or smother the flames with a coat or blanket. Do not throw water or shake the victim.

12.2. MATERIAL POISONING

If poisoning occurs through inhalation, take the victim to a ventilated area. When contact with the product is through the skin, wash the affected area with running water. Inform first responders of the material used in the accident.

12.3. BURIAL

Assess the site and check whether there is still a risk of further landslides. Only approach if there is no risk to yourself. Locate the victim if possible, removing debris from the nose, mouth and chest, combining the pressure being exerted.

Find out how many workers are there and inform the fire brigade.

12.4. FALLS FROM HEIGHTS

If a colleague has an accident and falls from a certain height, don't move him or her. Try to keep them awake and still, talking to them so they stay calm.

12.5. CONVERSATION

Keep the victim's head turned sideways, supported by your hand. Wait for the crisis to pass, not hindering their movements or trying to pull their tongue out.

12.6. PERFORATIONS AND HAEMORRHAGES

If the foreign body is small, like a nail, remove it and keep pressure on it to stop the bleeding. If the object is large, it shouldn't be removed, just put a cloth over it to stop the bleeding.

If the eye is punctured, don't touch the wound, just cover the victim's eyes to prevent movement.

12.7. ELECTRIC SHOCK

Do not touch the victim if they are still shocked, switch off the power supply. If the casualty is unconscious and not breathing, locate the middle of the chest, the tip of the sternum bone. Place the palm of your right hand two fingers above this point. Place the left hand on top of the right, with the fingers interlaced and raised, so that the right palm is prominent. With the arm at 90° to the victim, the rescuer will release the weight of their body onto the palm of the hand, applying compression, repeating the movement 30 times, once every second.

If the victim does not respond to the above procedures, push lightly on their chin and forehead with two fingers, as if they were going to look up. If necessary, apply the massage again, alternating between the two procedures, until the victim responds or help arrives.

BIBLIOGRAPHICAL REFERENCES

GASPANI, Claudia, Understand (once and for all) what a mentor can do for your career, 2017. Available at: http://exame.abril.com.br/carreira/entenda- de-vez-what-a-mentor-can-do-for-your-career/#

Guimarães, David B. O. et al. Health and safety in the construction industry: a report on the contributions of nursing. Revista de Enfermagem UFPE On Line, 2017.

KOPSCHITZ, Pedro, BUILDING CONSTRUCTION - UFJF,2011.

MESEGUER, A. G. Quality Control and Assurance in Construction.

Translation: Roberto Falcão Bauer, Antônio Carmona F., Paulo Roberto do Lago Helene. São Paulo: Sinduscon. Projeto/PW, 178p. 1991.

NETO, NESTOR WALDHELM, 2014 The importance of safety boots - DDS, https://segurancadotrabalhonwn.com/a- importancia-da-botina-de-**seguranca-dds/ acesso em setembro/ 2017**

NR, Regulatory Standard Ministry of Labour and Employment **NR 15 HARMFUL ACTIVITIES AND OPERATIONS 2014**

NR, Norma Regulamentadora Ministério do Trabalho e Emprego. NR-18 - Working Conditions and Environment in the Construction Industry. 2009.

NR, Norma Regulamentadora Ministério do Trabalho e Emprego. NR-26 - **SAFETY SIGNALLING** . 2011.

Oliveira, Ana Maria S. S. Construction and validation of a knowledge transfer model based on the training of construction workers, Florianópolis, 2010.

Profile of the Civil Construction Worker 2015 SINDUSCON Greater Florianópolis. 2015 and 2016

PLANALTO. Noticia. Disponívelem : http://www2.planalto.gov.br/acompanhe-planalto/noticias/2017/02/setor- da-construcao-civil-aposta-em-c crescimento-e-geracao-de-empregos- com-mancas-no-mcmv. Accessed: 28/08/2017

RAIS 2014 (Annual Social Information Report) http://www.rais.gov.br/sitio/retificacao/anterior.jsf

ROSSI, Fabrício, Main Tools Used on Construction Sites, Step by Step! Available at: http://pedreirao.com.br/principais- tools-used-on-worksites-step-by-step

Quality management - Guidelines for training NBR ISO 10015, 2001

Vitória, Daniele M. Evaluation of the Impact of Training at Work, UFB, Brasília, 2014.

ACKNOWLEDGEMENTS

To God for giving us the conditions to carry out the research;

To CNPQ (Internal Funding) - Federal Institute of Minas Gerais Santa Luzia Campus, to its administrative staff who provided the window so that today we can glimpse a higher horizon, filled with confidence in the merit and ethics of those present here;

To our supervisor, Wemerton Luiz Evangelista, for his support, corrections and encouragement;

And to everyone who helped bring this research to its significant conclusion.

Buy your books fast and straightforward online - at one of world's fastest growing online book stores! Environmentally sound due to Print-on-Demand technologies.

Buy your books online at
www.morebooks.shop

Kaufen Sie Ihre Bücher schnell und unkompliziert online – auf einer der am schnellsten wachsenden Buchhandelsplattformen weltweit! Dank Print-On-Demand umwelt- und ressourcenschonend produzi ert.

Bücher schneller online kaufen
www.morebooks.shop

Printed by Books on Demand GmbH, Norderstedt / Germany